OWNER AND SU

MW01113345

OWNER NAME _____

_____

HOME PHONE _____

WORK PHONE _____

CELL / MOBILE _____

E-MAIL _____

ADDRESS _____

_____

_____

SUBJECT / TITLE _____

_____

CLASS _____

INSTRUCTOR _____

START DATE _____

END DATE _____

VOLUME _____

NOTES _____

_____

Please contact the owner if you find this notebook. Thank you.

Graph Paper Notebook
1 cm Squares - 1 Square per cm
8.5" x 11" - 21.59 cm x 27.94 cm
200 Pages - 100 Sheets - White Paper - Page Numbers
Table of Contents - Owner / Subject Page
Softcover - Perfect Bound

We have a wide assortment of notebooks, journals, diaries, manuscript paper, calendars, planners, log books and other products. Choose from an extensive selection of cover designs, page sizes, page counts, paper, formats, styles and languages.

www.CactusPublishing.com

Published in Canada by Cactus Publishing Inc.

Author: Marc Cactus

Sample Page
Graph Paper Notebook
1 cm Squares

**This ruler is only shown on this sample page.
It does not appear on the actual graph paper.**

10
9
8
7
6
5
4
3
2
1
cm 0

cm 0 1 2 3 4 5 6 7 8 9 10

**This sample page shows the page number as iii.
The notebook pages are numbered from 1 to 200
and start after the table of contents.**

IMPORTANT NOTES

TABLE OF CONTENTS

DATE	DESCRIPTION / TOPIC / SUBJECT	PAGE

TABLE OF CONTENTS

DATE	DESCRIPTION / TOPIC / SUBJECT	PAGE

TABLE OF CONTENTS

DATE	DESCRIPTION / TOPIC / SUBJECT	PAGE

TABLE OF CONTENTS

DATE	DESCRIPTION / TOPIC / SUBJECT	PAGE

TABLE OF CONTENTS

DATE	DESCRIPTION / TOPIC / SUBJECT	PAGE

TABLE OF CONTENTS

DATE	DESCRIPTION / TOPIC / SUBJECT	PAGE

TABLE OF CONTENTS

DATE	DESCRIPTION / TOPIC / SUBJECT	PAGE

TABLE OF CONTENTS

DATE	DESCRIPTION / TOPIC / SUBJECT	PAGE

TOC-8

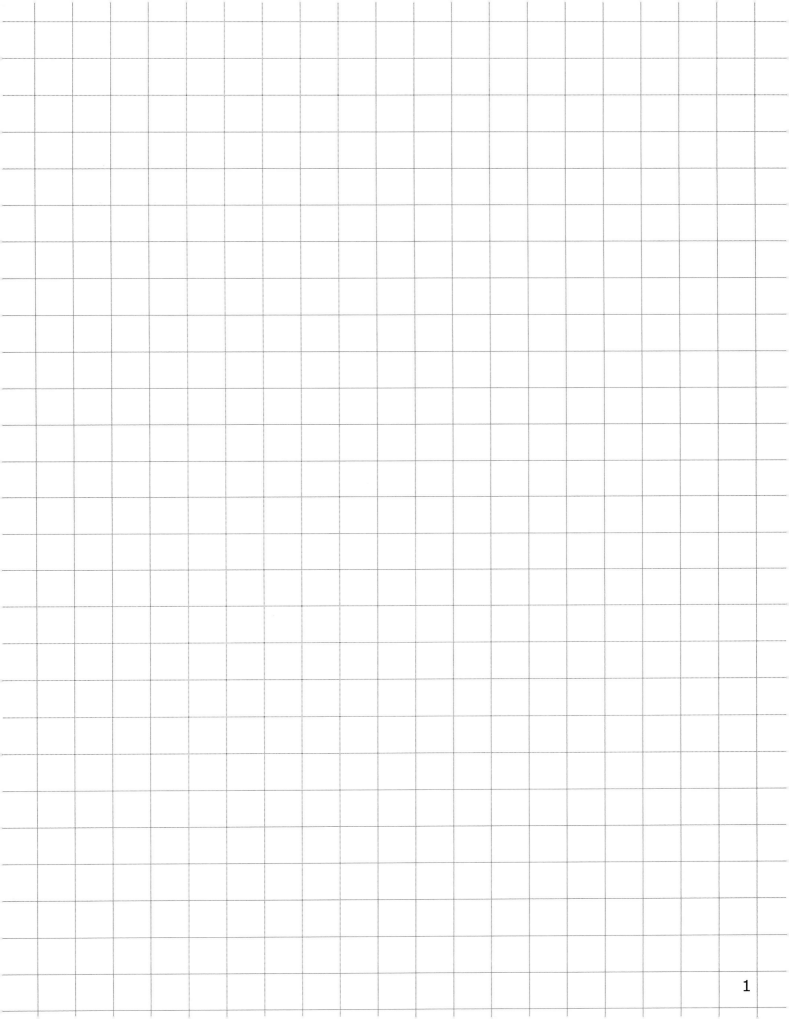

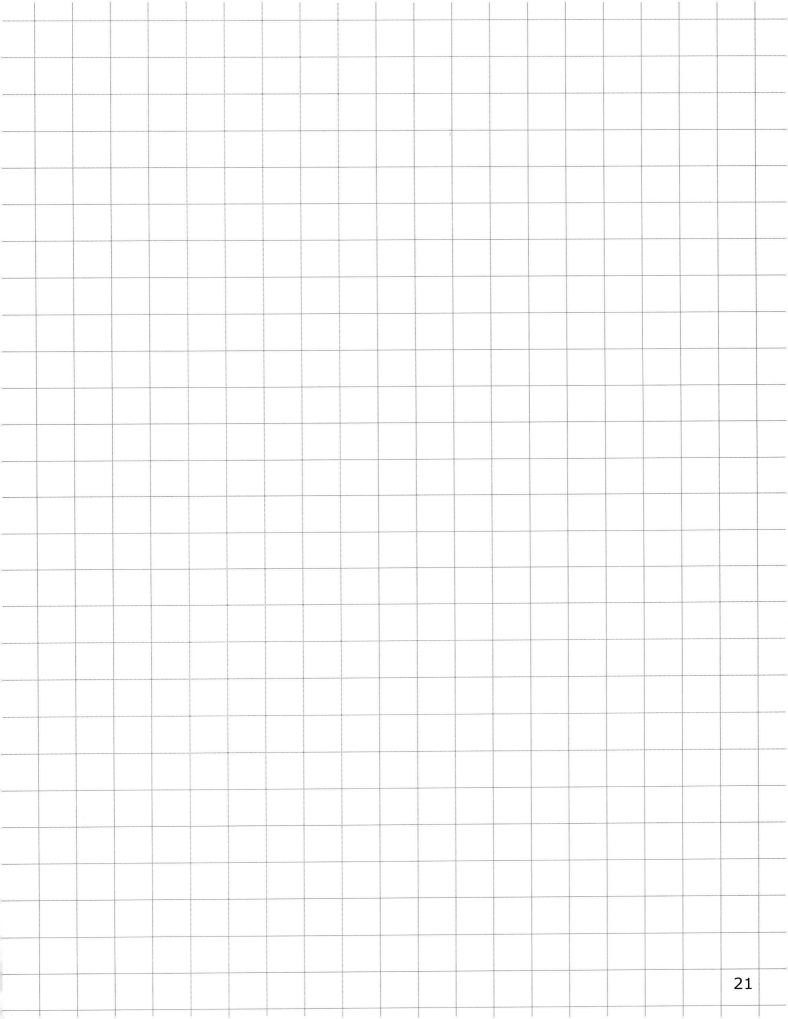

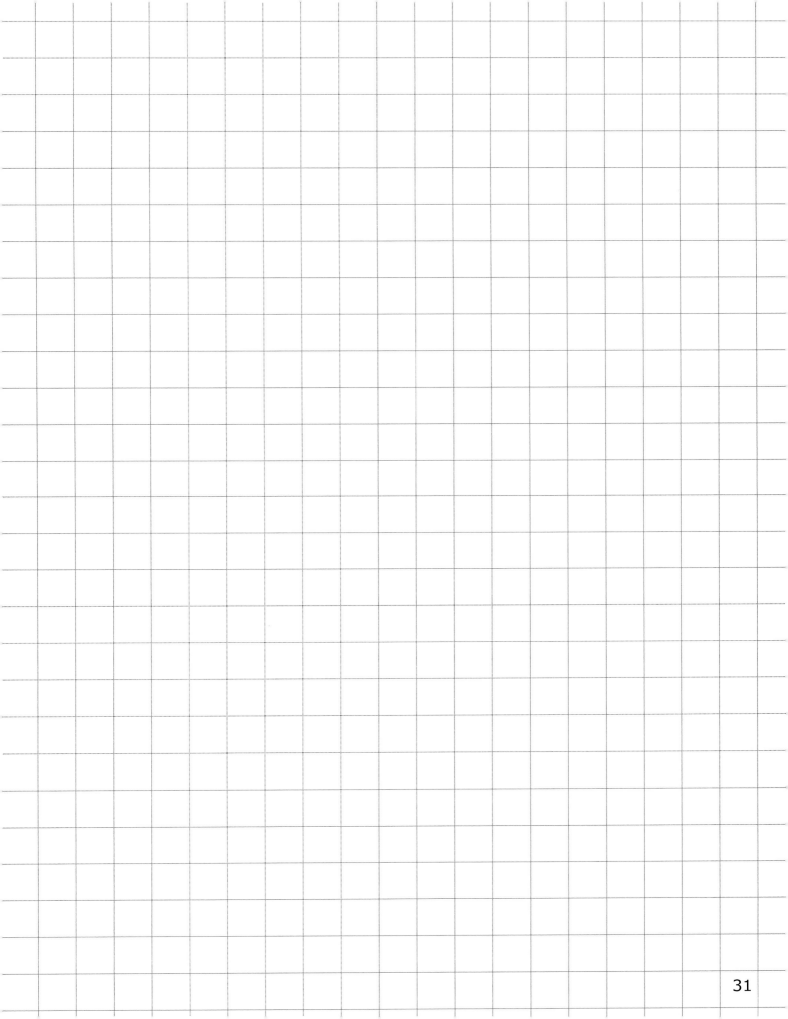

31

35

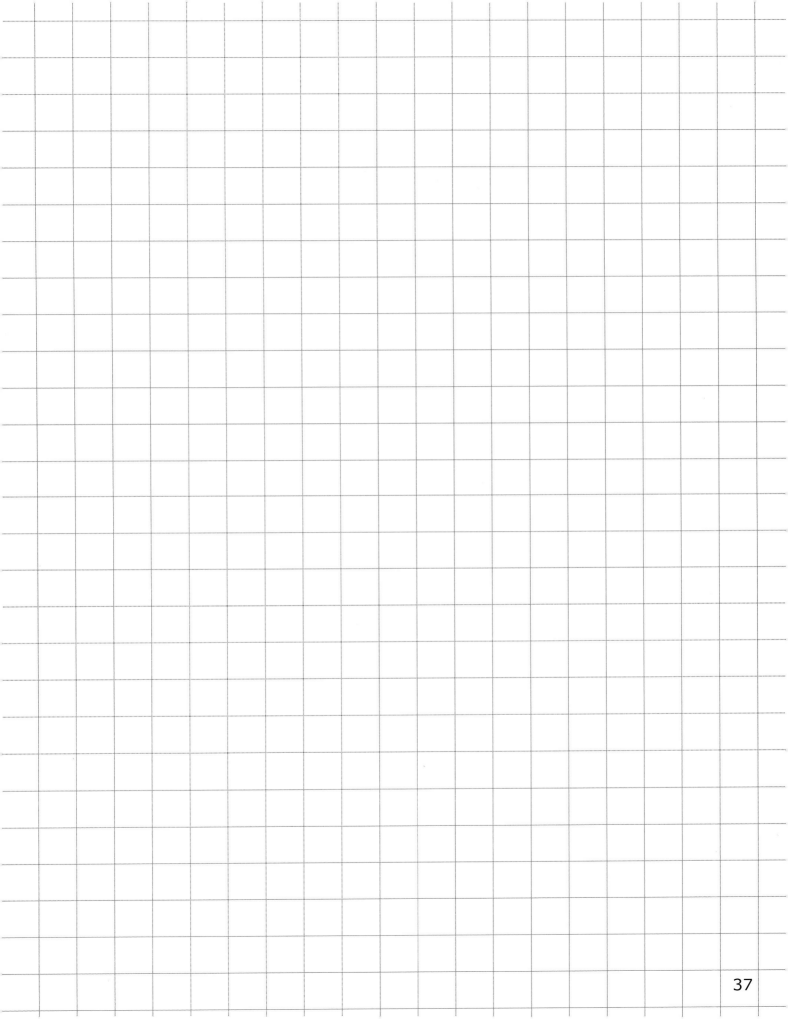

51

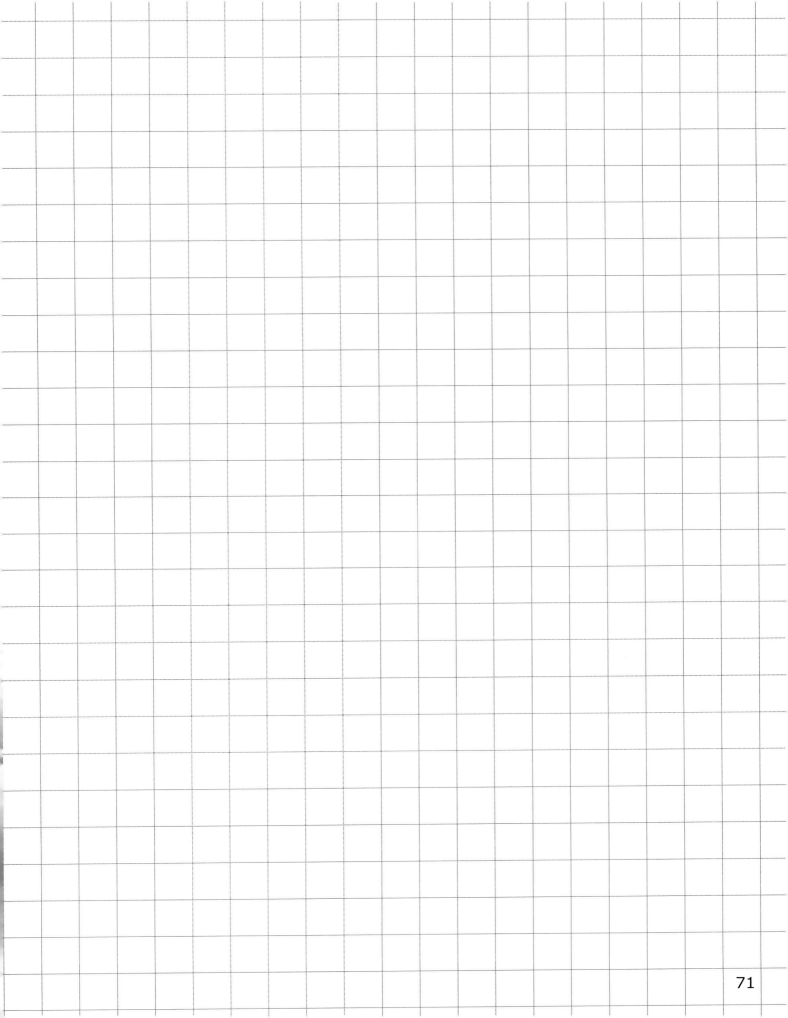

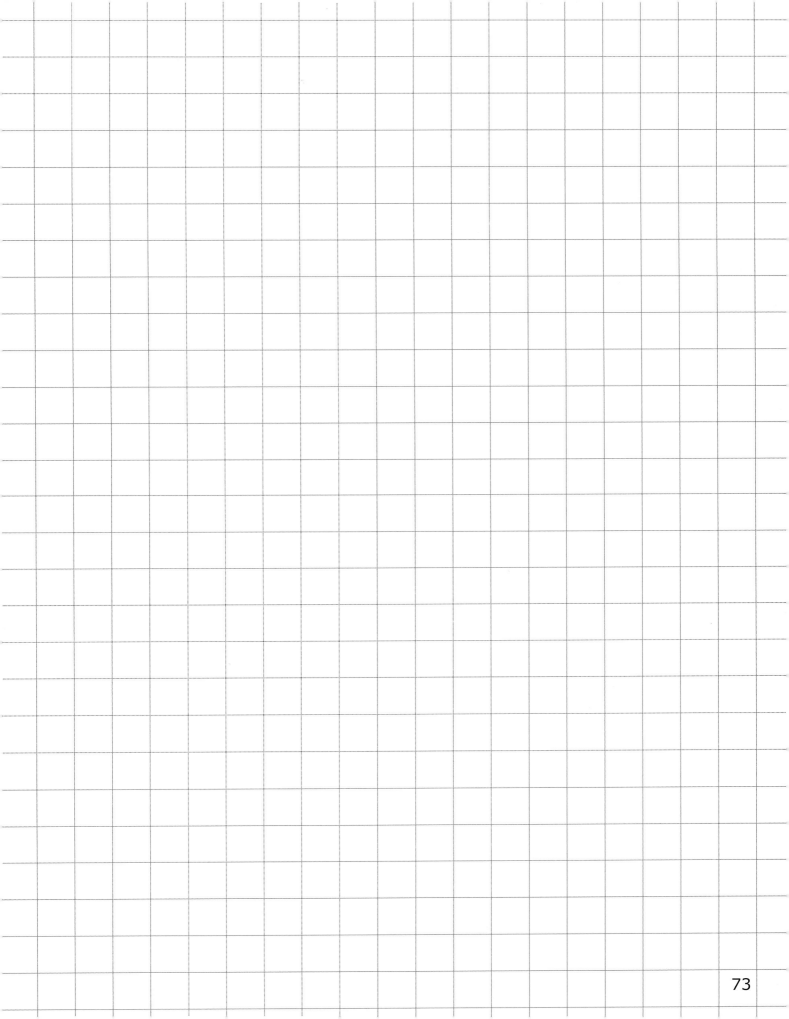

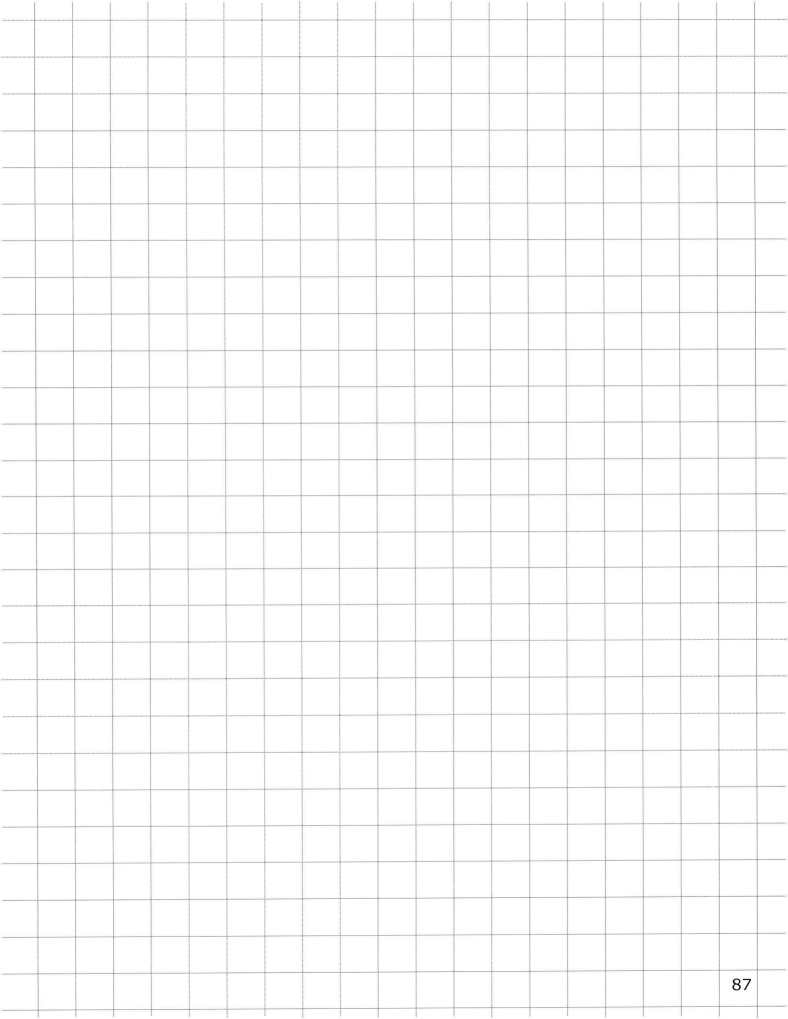

87

108

151

Made in the USA
Middletown, DE
16 August 2022